The Fish

T HE WATERS around the British Isles are very rich in fish, with a great variety of species being attracted by the wide range of environments – from the deep water and narrow rocky inlets of the Atlantic coast to the shallows of the North Sea and the Irish Sea – and by the great quantities of plankton which live in the warm currents of the Gulf Stream.

Of the species which frequent these waters, the four which dominated the catches of Scottish fishermen (before the advent of trawling) were ling, cod, haddock and herring.

Ling, cod and haddock are demersal fish: that is to say, they live on the ocean floor, eating smaller fish, crustaceans, worms and anything else which passes before them. From earliest times these fish have been caught on long lines with baited hooks; ling and cod – 'great fish', which can weigh up to 40lb – in the deeper waters, and the smaller haddock – never over 10lb – inshore. Herring, on the other hand, are pelagic fish, which move through the upper waters in huge shoals and feed on plankton. Such fish can only be caught by net.

The demersal fish are generally reliable in their movements, and can be found in particular locations at particular times of the year, year after year. This reliability, along with the simplicity of the boats and lines required to catch them, allowed many communities around the Scottish coast to supplement their income or diet with a regular, small, winter fishery from a time before records. This procedure remained attractive in some areas until the more efficient, and expensive, trawling methods superseded it in the 20th century, and put an end to this most ancient of traditional fisheries.

Line fishing, however, could never have fuelled the massive growth in the Scottish fishing industry which occurred in the 19th century. This was based upon the herring: an altogether more erratic and mysterious creature.

The herring is a small, silver fish, related to the pilchard of the warmer, more southerly waters. It reaches only a foot to a foot and a half in length, but travels in shoals which may reach several square miles in size. Like the demersal fish, the herring has an annual migration cycle, but it travels further and faster, and never remains in any district for more than a few days. Furthermore, it is always liable to change the pattern of its migration without warning, leaving whole communities without the means of livelihood. This occurred in the 14th century, when the herring forsook the Baltic, and again in the 20th century, when they gradually retreated from Loch Fyne.

Such changes of habit may have been due to alterations in climate or plankton levels, but to the fishermen who relied upon the arrival of the herring for their living they were entirely mysterious. It is not surprising that the herring fishermen were prone to both religion and superstition; generally making little distinction between the two.

Superstitions were so plentiful and localised, and their origins so obscure, that there is little point in listing them in detail. Who can guess why the fishermen of Tarbert, Loch Fyne, believed that a tiller made of rowan brought luck? Or why

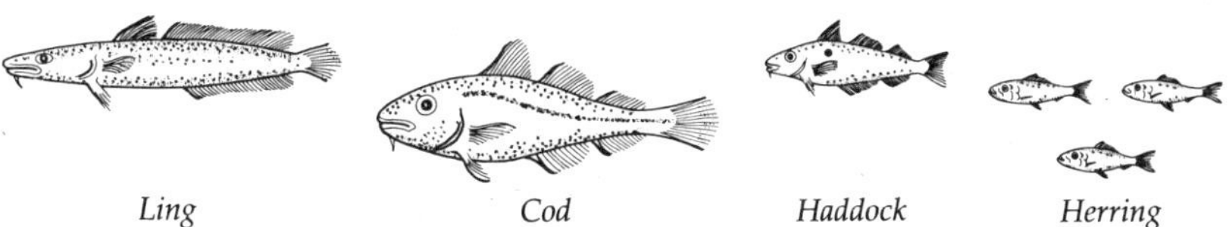

Ling   *Cod*   *Haddock*   *Herring*

the North East fishermen sprinkled their homes with salt water each New Year, and hung seaweed above their doors and hearths? Or why a chance meeting with a minister, or perhaps a red-haired person, would be considered so bad an omen that fishing would be abandoned for the day? Some customs, such as breaking a bottle across the bows of a new boat, still persist, but they are nothing like as numerous as they once were.

Religious beliefs had a similarly irrational, and rather feverish quality; periodically arriving as intense evangelical fashions which took a firm grip on some stretches of the coast, but completely missed others. In 1815 a law was passed forbidding the shooting or hauling of nets on the Sabbath within two leagues of the coast. The Tarbert ring-netters were quite happy to fish on a Sunday, and suffered no apparent ill-effects as a result, but the East Coast fishermen observed the rule strictly, and were prepared to lose out by so doing. For example, if a boat was delayed on its return from the fishing, and failed to reach harbour before Sunday, the crew would not land the catch, even although it meant that the fish would deteriorate to such a degree that they could not be sold. The failure of Dunbar as a fishing port was widely blamed on a failure to observe the Sabbath.

The unpredictability of the herring's movements was one of the frustrations of fishing for it, but it was also one of the attractions. For many years the average income from line-fishing was similar to that for herring fishing, but line-fishing lacked the opportunities for quick profits and gradually declined as a result. The uncertainties of the herring shoals gave the fishing, like gold mining, the allure of gambling.

However liable to alteration the herring's migratory cycles may have appeared locally, they did follow broad guidelines, with the herring shoals reaching different parts of the coast at different times of the year. In general, they were concentrated on the West Coast in the winter, in Shetland in the spring and early summer, off the East Coast in the late summer, and off the English East Coast in the autumn. As the industry developed, this staggered arrival of the shoals allowed the fishermen with larger vessels to follow the fish virtually full time.

The herring shoals which appeared along each stretch of coast were not comprised of the same fish, but of separate shoals with different migration cycles.

Morning catch of herring at Lowestoft

The Early Industry

T HE GREAT shoals which gathered off the coasts of Europe clearly presented an opportunity for profit, but the herring presented two difficulties for the early fishermen and merchants. Firstly, how to catch it, and secondly, how to keep such a highly perishable commodity in condition long enough to trade it.

It was the Dutch who first found the answers to these problems. In 1386 William van Beukels perfected a means of preserving herring, and shortly afterwards the first drift-nets were produced at Hoorne. With these two developments, and the advantage which they gave over any competition, the Dutch had the basis for a large-scale fishing industry which was to dominate the herring fishery for 300 years.

The drift-net was the first means devised for catching large quantities of pelagic fish. It is essentially a passive trap: a long wall of nets suspended from the surface of the water by a row of floats, with meshes of such a size that the fish swim into them and then become caught about the gills. Herring only tend to rise to the surface of the water at night, so the classic pattern of drift-netting is to set the nets at evening and haul them at dawn.

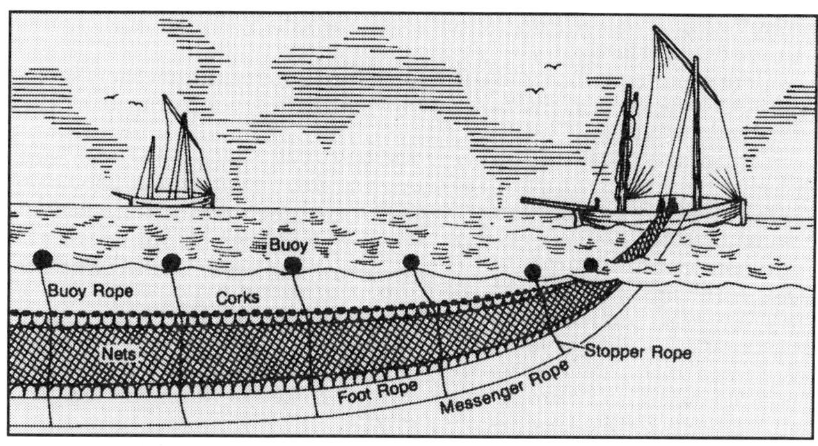

A herring lugger shooting a drift-net

Drift-nets remained the primary means of catching herring until the advent of the seine-net in the 20th century, but there was a considerable variation in the nature of the drift-net fishery in different places at different times, and the Dutch fishing industry was organised in a very different way to that developed by the Scots in the 19th century.

The Dutch fished from large ships called 'busses', in which they would follow the visiting herring shoals each year. They would start off the north of Scotland in the early summer, and would then gradually move south down the North Sea until the season ended off the south east of England around Christmas. The nets were set from the busses, and each day's catch was cured and barrelled on deck, then picked up by one of the fleet of small, fast 'jagers', which serviced the fleet throughout the season and delivered the catch to British or European ports for distribution.

At the height of the Dutch industry, in the 17th century, over 2000 boats were involved in the North Sea fishing, and Dutch cured herring were unrivalled throughout Europe. Naturally, the other North Sea countries were jealous of the success of the Dutch, and the revenue which they generated from the herring; particularly the Scots and the English, who watched the fishing take place off their shores each year.

Attempts were made by both Scotland and England to limit the Dutch fishing, but these were generally rendered ineffective by a lack of naval power to enforce them, and the monopoly was not broken until the War of the Spanish Succession (1702-13), during which the Dutch lost their markets. After this time, although the Dutch retained a high reputation for their cured herring, their fleets tended to decline in the face of new competition from countries such as Scotland and Norway which were nearer the major fishing grounds, and which were pioneering a much more efficient and flexible system.

Scotland's early attempts to harvest what was one of its prime natural resources were hampered by a number of factors; political, legal and economic. Scotland remained a very poor country until the 18th century. Despite being part of the United Kingdom, its merchants were excluded from trading with England's burgeoning New World colonies until after the Union of Parliaments, in 1707. Even after that, Scotland remained in a state of political upheaval until after 1746, when the last of the Jacobite uprisings was quelled and the pacification of the Highlands could be completed.

More specifically, Scotland's fishing industry suffered from short-sighted legislation which effectively thwarted any attempts at development until the end of the 18th century.

The Scottish coast – particularly the east coast – was lined with small communities which made part or all of their income from small-scale line-fishing in open boats. The people in these communities only needed encouragement and the injection of a little money to form the backbone of a new herring fleet, but the legislators were handicapped by a mistaken belief that the Dutch fishing method could not be improved upon. Thus, when bounties were first introduced, in 1718, they were paid not on fish caught, but on the tonnage of boats built – specifically busses – and it was prohibited for these vessels to buy herring caught by smaller boats. In addition, the number of ports which could handle fish was severely restricted by law, and a prohibitive duty was charged on the import of foreign salt, necessary for the herring-curing process. In principal, this duty did not apply to fish-curers, but in practice only the largest could avoid it.

The cumulative effect of these laws was to exclude the poor fishermen from the herring industry, and to keep the development of fishing in the hands of a small number of fishing companies, established by merchants, which could afford to build a fleet of busses; with (as the economist Adam Smith observed) the 'sole purpose of catching, not the fish, but the bounty'. Busses were built, but their combined efforts appear anaemic in comparison with the explosion of fish landings which was to come, and once small-boat fishing had been established they quickly disappeared.

At the end of the 18th century the restrictions were gradually lifted, and, most importantly, the bounty was shifted to barrels of herring caught by small boats. The brakes had been removed, and the coastal communities stood on the brink of an unprecedented prosperity. The herring fishery would continue to grow for over a century.

Caithness

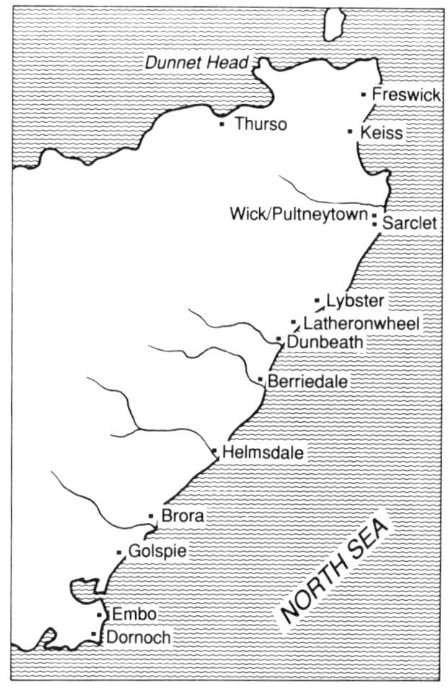

FOLLOWING THE Jacobite defeat in 1746 there was a sad period of confusion and change in the Highlands. The clan system and the way of life which it encouraged were clearly no longer sustainable, and Highland landowners sought for ways of making their estates more profitable. The net result of these changes was the large-scale movement, forcible or otherwise, of large numbers of Highlanders from their homes. Some settled on new, less fertile, crofting settlements along the coasts, while others went further, to the growing industrial towns to the south or the new lands of the expanding empire.

The social upheaval of this period inspired the creation of a number of societies, dedicated to preserving as much as possible of the threatened Highland culture. Some of these societies went further, and sought ways of restabilising the Highland population by encouraging agriculture and industry.

The Highland Society of London was formed in 1778, and its members took a keen interest in the economic development of the area. In 1786 John Knox – a London bookseller who had travelled extensively through the Highlands – gave a lecture to the society on the desirability of establishing a number of villages along the west and north coasts to act as both a base and a market for a Highland fishing industry. In the same year a committee of members of the society transformed themselves into a joint stock company called the 'British Fisheries Society' and began to raise money in order to pursue Knox's policy.

Previous fishing companies had concentrated on building fleets of busses to profit from the bounties and fish sales, but the BFS chose to pursue a different policy; planning to provide the land, buildings and harbour and to leave the fishermen to look after the fishing themselves.

Knox had recommended the building of 40 villages between Arran and Dornoch, but in the event only four were started: Ullapool, Tobermory, Lochbay and Pulteneytown.

The villages met with varying degrees of success. Lochbay, on Skye, never developed beyond a small crofting community; while Tobermory, by a fine natural harbour on the island of Mull, was too far from the fishing grounds to develop as a fishing port, though it quickly became, as it remains today, the principal local service centre. Ullapool was initially more successful, but soon declined in importance, and has only recently recovered as a fishing centre; notably as a base for Eastern European factory ships. The most successful of the villages was Pulteneytown (now a part of Wick, which was then a tiny crofting township), on the eastern coast of Caithness. The village was founded in 1808, and by the mid-century was one of the most important herring ports in Europe.

Pulteneytown succeeded where the other villages failed for a number of reasons. Firstly, it was nearer a major fishing ground than the West Coast villages. There had been a small herring fishery off Wick as early as 1767, which, in addition to the winter line-fishing, had encouraged a number of men to look on fishing as their primary occupation. This made the transition to full-time fishing easier for the people of Caithness than it was for the West Highlanders, who considered themselves primarily to be crofters.

In addition, Pulteneytown, and the Caithness coast in general, was much more successful in attracting money than the West Coast; both from central government and, more importantly, from the curers.

The curers were the key to the success or failure of herring fishing in any given location. The herring might be plentiful, but if there were no curers they could not be sold, and if they could not be sold they might as well not be there. The curers dictated when, where and for how long the herring fishery was pursued each year.

When the herring bounty was transferred to barrels of cured fish at the end of the 18th century, established curers, notably from the southern coast of the Moray firth, moved to Wick/Pulteneytown and the other settlements along the eastern coast of Caithness and Sutherland and encouraged the local fishermen to work for them in the fishery. In this they faced stern competition from local, monied men, who quickly set up as curers when they saw the profits to be made.

The system worked like this. The fishermen were poor and could not afford the quality of boats required for efficient drift-netting, therefore the curers either advanced them the money to buy such a boat, or took a share in the boat themselves. In return, they made an agreement to take so many crans (the established measure for landed herring: 37½ gallons) at a set price, in addition to paying an agreed bounty, a spirits allowance to the crews, and 'arles': a sum paid at the end of each season to guarantee the use of the boat the following year.

Wick harbour crowded with open fishing boats, 1860s

Lybster harbour at the height of the herring season

In the early days of the Caithness herring fishery the crews had little difficulty in paying off their debts to the curers, but the continuing need to update equipment ensured that the curers retained a strong direct financial interest in the fishing fleet.

The local market for fish was tiny, so once the herring had been landed, they were cured and exported; some to the industrial cities to the south, but an increasing proportion overseas. At first, the major overseas markets were Ireland and the West Indies (where the herring were used to feed the slaves), but in 1815 the first Scottish barrelled herring were exported to the continent, and by 1850 Europe took the majority of Scots exports.

The curers would arrange for ships to collect the barrels and transport them across the North Sea; the majority going to the ports of Hamburg, Stettin and Danzig (then all in Germany, though the last two are now in Poland) from where agents distributed them throughout Germany, Poland and Russia. The relationship between the agents and the curers was similar to that between the curers and the fishermen, and the curers' loans to the fishermen often originated as a loan or advance from an agent to a curer. Thus, the whole industry ran on borrowed money, and a string of jobs could be dependant on a steady income from each brief fishing season. Unfortunately, the herring is not a reliable fish, and the inevitable bad years would result in a rash of bankruptcies and closures. Never enough to halt the growth of the industry, however, for if the supply was erratic the demand was not, and there were always new men prepared to gamble for the potential high profits.

In the early years of the Caithness herring boom, there were only two harbours along the coast: the BFS harbour at Wick/Pulteneytown and a smaller one at Lybster, to the south. The lack of harbour space was a problem for the whole herring industry during this boom period, and put a severe limitation on the size of boats which could be used, since they generally had to be hauled up the beach by the crew after use.

After the establishment of Pulteneytown, almost all the BFS's expenditure on the site was on the harbour. By 1834 they had spent about £30,000 (a considerable amount at a time when a new 40' fishing boat would have cost about £60 in total), but large numbers of boats, even at this major centre, were still obliged to use the foreshore and the nearby mouth of the River Wick.

To put the problem in perspective: from a cold start at the end of the 18th century the summer herring fleet which gathered at Wick had risen to around 500 boats by the early years of the 19th century, and to 800 by 1848. By the 1860s it was over 1100, with the size of the boats steadily increasing.

This kind of overcrowding was repeated at all the available harbours on either side of the Moray Firth, and the potential dangers were tragically realised in the great storm of 18th Aug, 1848, when 41 of the 800 boats which set out from Wick were lost, and 25 men with them; mostly trying to make the narrow entrance of the crowded harbour. 100 men and 124 boats were lost altogether from the coast between Wick and Peterhead.

A thorough enquiry was held into the disaster, and a review compiled which harshly criticised the types of fishing vessel used (Scottish boats were still undecked at that time), the quality of the harbours, and the competence of the harbour administrators. It was made clear that on that unhappy night even the harbour space available was effectively useless because the entrances were blocked by trading ships and, in one case, old masonry. Furthermore, the lack of adequate harbour lights, or even ropes to throw to the fishermen whose boats were wrecked on the harbour wall, caused the unnecessary deaths of a number of men within sight of home.

The result of this investigation was a gradual improvement in both boat design and harbour management, but it was always a struggle to keep up with the swift expansion of the industry; not simply in the fishing facilities, but also in the housing arrangements for visiting crews and migrant, seasonal workers from the West Highlands. These problems were not helped by the tendency of the industry to concentrate on a smaller number of larger harbours as the size of boats gradually increased.

This was not such a problem in the early days of the industry. With the exception of the Wick fishermen, the men who fuelled the first expansion of the fishing were still primarily crofters, who took part in the herring fishing for no more than two or three months of the year, between June and September: the time of year when the herring were in their immediate vicinity. Their boats were no more than 20' in length, and since they were so small they could be operated from the open shore; and since one shore was as good as another, the fishing bases were small and numerous; tucked away in every little bay and creek along the coast, sometimes with the fishermen's crofts at the top of a high cliff behind the narrow foreshore.

Whether at Wick or Lybster, or at one of the smaller stations, the fishing and curing routine was broadly similar, and provided a blueprint which was to be copied by other areas of Scotland as the fishing spread.

In July the visiting curers would arrive, while those who lived on the coast would stock up with salt and barrel staves. Each evening the fleet would sail out to the fishing grounds and set their nets, and the men would find what comfort they could in the small, open boats. At dawn the nets would be hauled and the fleet would return to the shore. Successful crews would land their catch while the curer stood by, checking the number of crans before having the fish carried to the gutting benches.

For every boat hired, the curer would take on a crew of three gutters (they

Fish-gutting at Castlebay, Barra

would not work on the catch of a single boat; that was simply the ratio needed). The gutters were women, very often from the West Highlands and the Western Isles, and competent workers were prized. A good crew might be paid arles and tied to a single curer just like a boat crew. The work was hard, however, and during the height of the fishing the herring girls might be required to work from six in the morning till after midnight.

The crew was split into two gutters and a packer. The gutters would gut the fish with a single deft movement of a knife, then the packer would salt them and arrange them in the barrels. Each barrel contained around 700 fish, but such was the speed of the herring girls that it might take only ten minutes to fill.

The first job each morning was to top up the barrels filled on previous days. Each barrel was left open for a period of 15 days, and during this time the fish would slowly settle down into the brine, leaving a slight ullage each morning.

At the end of the 15 days, the barrels were checked by a Fishery Board inspector. If the barrels were full and showed no sign of leakages, and the herring had been correctly cured, they were sealed up and branded.

There was a small charge for the government brand, but since it was an accepted standard throughout Europe, and a barrel with a good brand could expect to receive a price up to 50% higher than one with an inferior brand, and even an inferior brand would raise more than no brand at all, the majority of barrels received one, though it was not compulsory.

Barrels were branded according to the quality of the fish; the variation in quality reflecting the stage the fish were at in their spawning cycle.

The main categories were Mattie, Full, Large Full and Spent. Matties were fat fish, caught before their roe or milt had developed. In Fulls the roe and milt were developed, while Spents were those fish caught after spawning, and were the least valuable of the three types.

After barrelling the herring would be loaded on to trading ships and exported as swiftly as possible. The herring market was prone to cycles of gluts and shortages which could cause great variations in prices, therefore it was vital for

Wick harbour. Decked luggers with their sails set.

the curers and their agents that the fish should not be late in reaching their potential customers.

The fishing would continue for as long as the herring remained in the district, or for as long as the curers wished it to. When it finished, the crofter/fishermen would haul up their boats and return to working the land, while the curers and full-time fishermen would look for ways of extending the season. Many were involved in the winter line-fishing, but having developed a taste for the excitement and potentially high profits of the herring fishing, they naturally wished to exploit the shoals which reached other parts of the coast at different times of the year, and which were still largely unfished.

Initially, the Caithness boats were too small to make long voyages, particularly across the rough waters around Northern Scotland. By the 1840s, however, the larger, 30'-40' boats began to make the journey through the perilous Pentland Firth to the Western Isles. The crews would fish the shoals which gathered in the Minch in May-June, then return to their home bases in time for the Caithness fishery in July.

In the 1880s an alternative May-September fishery was opened up in Shetland, and by the end of the century Caithness boats were travelling south for the autumn fishery off East Anglia. By this time the Scottish sailing fishermen were at their height, and the fleet which sailed south included boats based in Shetland and the Western Isles, as well as those from Caithness and the older fishing communities down the rest of the East Coast.

Barrel brands

The East Coast

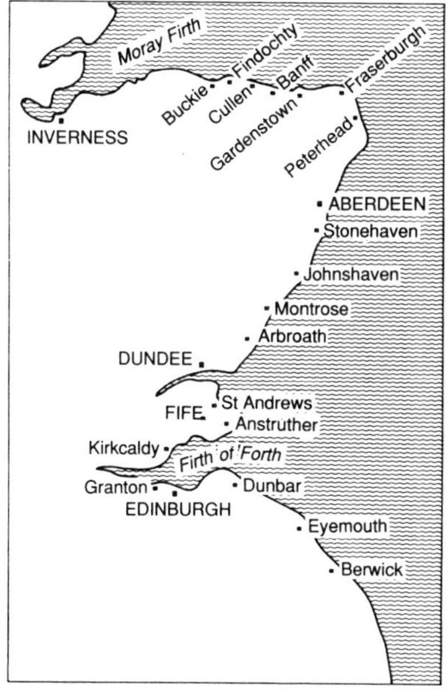

THE NEW herring industry was largely confined to the east coasts of Caithness and Sutherland until about 1815, but older fishing communities existed elsewhere in Scotland before this time. Indeed, while most of the North's fishing communities dated back little further than the end of the 18th century, a large number of small communities on the remainder of the Scottish East Coast, between Inverness and Berwick, had been relying on fishing for the major part of their income since the 16th century.

The development of the herring fishing was slower along this coast, largely because the fishermen were already involved in profitable, if limited, line-fishing. As a result, they were less inclined to make sudden changes than the poor men of the far north; many of whom moved straight from crofting to herring-drifting, without any previous knowledge of fishing. Furthermore, the East Coast fishing communities were very clannish and independent, and crews were less inclined to take the curers' assistance in purchasing a suitable boat, preferring to save up painstakingly and buy one amongst themselves.

This is not to say that there was no herring fishing undertaken along the East Coast, but the shoals were not as reliable as they were further north. In the late 1700s the herring had appeared in great numbers in the Firth of Forth, and were sold fresh to the growing populations in Fife and the Lothians, but after 1805 they virtually disappeared. Each summer, some 80-90 Forth herring boats travelled north to Wick, sometimes rowing the entire distance, in pursuit of the erratic shoals.

The staple occupation of the East Coast fishermen was line-fishing: small-line fishing for haddock in the winter months, and great-lining for cod and ling between March and June.

The rewards for line fishing were regular but low, and the work was labour intensive, requiring the fishing families to work closely together in a long-established routine. The broader communities were linked together by the network of shares in boat ownership. One fisherman might own shares in a small haddock boat, a long-line boat of 25' or more, and even a herring lugger; each vessel being joint owned with a different crew.

These close-knit relationships tended to exclude outsiders, and the fishing families lived in their own small villages by the shore side, or, if in a larger town, in a recognised quarter, generally termed 'seatown' of 'fishertown'. This exclusivity was continually reinforced by the need for the young men to marry girls from a similar background, since, it was generally acknowledged, only someone bred to the work could stand it.

Line fishermen being carried ashore by their wives; Portessie (Buckie)

When the boys reached the age of 18 they took a share in a boat, and with it the responsibilities associated. These required the presence of a wife, and ideally a few children too.

In the small-line fishing, for example, though the men undertook the fishing during the day, the business of baiting the lines (generally with mussels) was undertaken by the wives and children. Since there could be up to 3000 hooks on a boat's lines, this was quite an undertaking.

In great-line fishing, the men baited the hooks themselves with haddock or saithe, on the way out to the fishing grounds, but the women had other responsibilities. These included selling fresh fish in the immediate neighbourhood, cleaning and curing the fish for export, helping to haul up the boats, and even carrying their husbands out to the boats, so they shouldn't start the trip with wet feet.

This close family interrelationship was gradually weakened by the rise of the herring industry, though it continued in some areas until the decline of line-fishing in the 20th century.

Initially, drift-netting for herring was just one in a range of fisheries which the East Coast fishermen pursued each year. The families would make the nets (until the introduction of net-making machinery around 1820), and the fishermen would take their boats to one of the stations where the curers had established themselves: initially in Caithness, but soon along the East Coast, particularly in the North East, with the ports of Fraserburgh, Peterhead and Aberdeen soon emerging as the most important (as they remain to this day). Their wives and daughters would follow to find employment as gutters, and often whole villages would be empty for the two to three months of the herring fishing. Following the end of the herring season, the families would return to their home bases, haul the herring drifters to the top of the beach, and return to the line-fishing.

Gradually, however, as the herring season extended and the boats stayed away for longer periods, the fishermen abandoned the line-fishing. At the same time, the continuing increase in the size of the herring luggers made it impossible to continue operating them from the open beach, with the result that many smaller communities had to either abandon herring fishing or else abandon their villages.

A number of villages did disappear at this time, but surprisingly few in the circumstances, and the East Coast – particularly in the North East – is still fringed with old settlements: small clusters of cottages, their gables presented to the sea, squeezed onto narrow foreshores or into narrow gaps in the high cliffs which predominate along this coast.

Most of these villages have lost any connection with the fishing industry, but some still retain large populations of fishermen, who choose to live in their traditional settlements and commute to the larger ports for their work.

This loyalty to their home villages was very marked in the East Coast fishermen, and, as long as there was a harbour large enough to take the fishing luggers, the peripatetic nature of the herring industry allowed them to fish from any base without penalty. This tendency was most marked in towns like Buckie, in Moray. The port had little or no home fishing, but held a huge stake in the herring industry. By the end of the century a large fleet of top-rate Buckie drifters vigorously pursued the shoals, only returning to their home base for a short period each year.

With their large number of professional fishermen and good seamen, comparative proximity to the home markets, and sound financial base, the East Coast fishermen made a great success of the herring fishery, and were responsible for most of the innovations in boat and equipment design, and in the opening up of new areas, from the mid 1800s onwards.

This is not to say they did not suffer reverses. The nature of the industry made commercial set-backs inevitable, while, on a more personal level, there were always likely to be losses of men and boats in the harsh environment of the North Sea.

In particular, there were heavy losses in the two great fishing tragedies of the 19th century. In the 1848 storm (already mentioned) the 300 strong fleet from

Baiting lines at Auchmithie, near Arbroath

Undecked luggers in Anstruther harbour, 1873

Peterhead lost 31 men and 51 boats. An even greater tragedy struck the Berwickshire coast on the 14th Oct, 1881, when, from a fleet of 45 boats, only 26 managed to return to the shore. A total of 189 men were lost, 129 of whom were from the single port of Eyemouth, which lost half of its fishermen in this one storm.

In general, however, the fishing industry progressed, and although the East Coast fishermen were as keen to stick to tradition as any, they were more inclined to face commercial facts, and to accept any changes which would benefit business.

For example, the advent of trawling for white fish (those previously caught by line) was greatly against the wishes of the herring fishermen, who believed that dragging nets along the sea floor destroyed the herring spawn which gathered there. The technique was clearly efficient, however, and as early as 1882 experiments were being carried out in steam trawling from Aberdeen harbour. In contrast, the first trawler catch was not landed in Wick until 1902, and the skipper of the boat was stoned for his troubles.

It is no coincidence that most of the major Scottish fishing ports are now on the East Coast.

Shooting a great-line

The West Coast

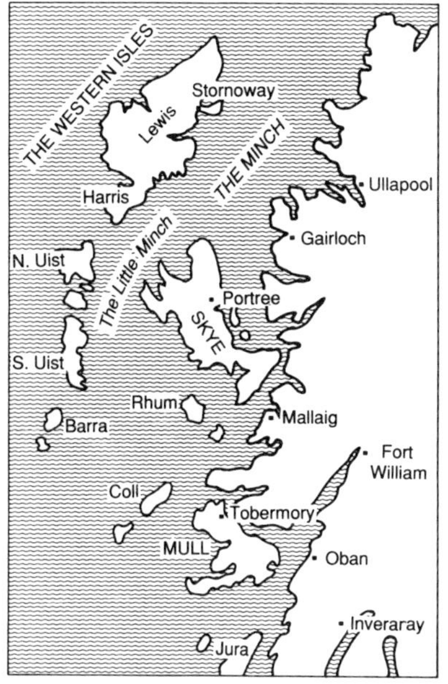

WHEN THE British Fisheries Society was founded in 1786, its major concern was to encourage the fishing industry in the Highlands. Of the four villages which it established, three were on the West Coast, and one, Pulteneytown, was on the eastern coast of Caithness. The major thrust of the society's endeavours can therefore be seen to have been in the West, yet of the four bases, only Pulteneytown prospered to any degree. The North Western mainland never developed an indigenous herring fishery of any importance, and the Western Isles, although they were to profit from the industry, were largely second-hand beneficiaries of the annual East Coast migration to the Minch, and the island fishermen never achieved the levels of wealth and independence reached in the East.

The reasons for this comparative lack of success are complicated, but they boil down to three major problems: communication difficulties, poverty and the lack of reliable inshore shoals.

The Highlanders were not strangers to fishing, but their boats and equipment were of such an inferior quality that they made little attempt to catch more than was required for simple subsistence. This need not have stopped them developing a line-fishing industry, had they not also been handicapped by problems of market and communications. Whereas the early East Coast fishermen were able to sell fresh fish to a relatively prosperous agricultural hinterland, no such market existed for the Highlanders; and while cured cod and ling could be easily exported from the East, the long distances and lack of regular services made such a thing difficult for the Highlanders. At the end of the 18th century, for example, the crofters of the island of Barra would send their catch of cured fish to the Clyde once a year; carrying it in their open fishing boats. In 1783 five boats made the trip, but four of them were lost with their crews on the dangerous crossing.

When the herring boom started, the West Highlanders should have been in a strong position to prosper, but they lacked the ingredient required to make the fishery work: the curers.

The curers were not unaware of the herring in the Minch. As early as the mid-18th century Clyde boats had been fishing the area in a limited way, with smacks rigged out for curing sailing north, and either taking two or three Clyde fishing boats with them to catch the fish, or buying fish from the small local boats. The problem was that the Highlanders' boats were not seaworthy enough to fish in the open Minch, and the herring did not regularly visit any of the lochs which deeply indent the mainland coast, so it was never easy to find them. Some years a loch would have no fish; others it would have a useless glut. The Highlanders

Luggers passing Kisimul Castle as they leave Castlebay, Barra

had no curing equipment, so there was no point in catching the fish unless the curing smacks were in the area. No industry of any scale could be based on the slim hope that the herring and the curers would be simultaneously present in any particular loch.

The Western Isles were nearer the main fishing grounds, but they could not be developed until the curers had first perfected the small boat herring fishery on the more accessible Caithness coast.

In the early days of the Caithness boom, the West Highlanders and Islanders were unable to profit directly (as the Forth fishermen were) because their boats were so inferior. However, many of them crossed the country each year to find jobs. In 1845 some 10,000 workers sought work in Wick alone, and the majority of these were Highlanders. The men vied for positions as hired deck-hands, or in the curing yards, while the women and girls found work as gutters.

The income from this work was meagre in comparison with that of the boat owners, but there were more people than jobs, and with a rising Highland population creating famine conditions in some years, the Highlanders had to take what work they could.

In the 1840s, with the value of herring continuing to rise, the full-time fishermen of the East began to look for a way to extend the fishing season, and decided to experiment with the early fishing in the Minch, in May and June.

The curers ignored the mainland and chose instead to operate from the Western Isles. At first they concentrated their efforts on the existing port of Stornoway, in Lewis, but soon (in an attempt to gain an advantage, however slight, over their rivals) they began to establish their own bases along the eastern coast of the islands. By the 1870s there were 16 separate bases established in the islands for the six week season.

These bases were not situated in existing villages, but in quiet bays and creeks, and their construction must have required a great deal of planning and organisation, since the whole station – jetty, curing facilities and accommodation for the gutters, porters and coopers – had to be pre-fabricated and assembled on the site.

At the peak of the Minch herring fishery – by which time sites had spread throughout the islands, and a large part of the fleet operated from Castlebay, in Barra – there were 1400 boats involved in the fishing. Although around 1100 of these were from Caithness and the East Coast (about a third of the fleet in those areas), the presence of the curers had also encouraged the establishment of a local fleet. Thenceforth, herring luggers from the Outer Isles began to compete not only in the Minch, but also in the Caithness and North East fisheries – an innovation not always appreciated in the East, where large-scale battles would occasionally break out between Gaelic-speaking islanders and their hosts.

The Island fleet continued to be an important element of the herring industry until the end of the century, and Stornoway and Barra boats would even make the long trip to East Anglia for the autumn fishery, but there was an essential weakness in the structure of the island industry which eventually caused its failure.

While the East Coast fishermen tended to buy boats from their own savings, the Islanders, being poor, had to follow the example of the Caithness fishermen and borrow from the curers. Unlike the Caithness men, however, they were rarely able to pay off their debts, and their income each year would often do little more than pay off that year's borrowings (principally for equipment; also supplied by the curers) leaving the standing debt untouched.

The reason for this was that the Island fleet was largely comprised of boats and equipment purchased second-hand from the East. By the 1840s the Scottish industry was continually gearing-up, with each year's boats becoming larger, and thus more productive and efficient, than those in use before. As a result, the Islanders were always operating with equipment which was slightly less productive than the market demanded, and though they were able to stay in operation, they could never prosper.

At the end of the century, the Islanders were operating with the largest, finest sailing luggers ever built, but by that time the industry was transferring to steam. Financially unable to make this expensive leap, the Islanders continued to fish their luggers until they became unseaworthy, then largely abandoned drift-netting.

Packing herring at Campbeltown

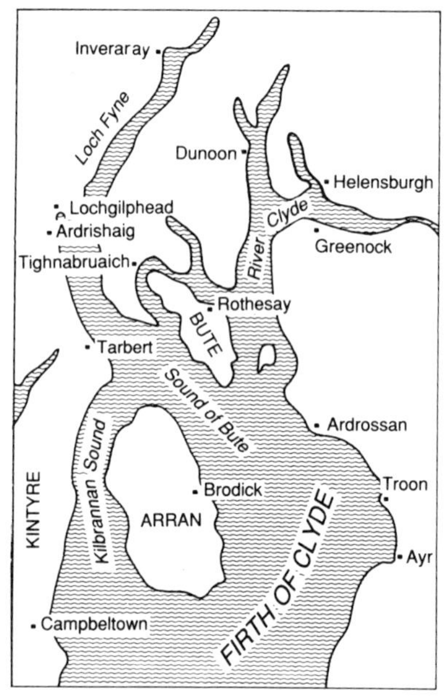

ALTHOUGH THE herring stocks were unreliable in almost all the lochs of the West Coast of the mainland, they did visit the Firth of Clyde each year in large numbers, and were particularly plentiful in the long, narrow inlet of Loch Fyne. In this area a community of innovative full-time fishermen emerged, whose industry and boats both developed along different lines to those of the rest of the Scottish coast.

Apart from the regular arrival of the herring, the Loch Fyne fishermen had a number of other advantages. Firstly, they fished in relatively enclosed waters, and were thus able to use smaller boats than those who fished in the open sea; a factor which greatly reduced their financial outlay. Secondly, they were handily placed for one of the fastest growing markets in Europe. Between 1708 and 1921 the population of Glasgow rose from 13,000 to 1,000,000 (not including the surrounding towns), and there were few families who did not regularly eat the cheap and nourishing Loch Fyne herring.

In addition, the Loch Fyne men developed a new style of fishing – ring-netting – which was both cheaper and more productive than drift-netting.

Initially, the Loch Fyne fishery was pursued much as elsewhere, with the herring caught by busses, and the crofter/fishermen augmenting their income with line-fishing for haddock, cod and ling. When the rules of the herring bounty were altered at the end of the 18th century, the Loch Fyne men quickly took to drift-netting from small boats and generally prospered from it.

In the 1830s, some of the poorer men along the loch began to experiment with trapping the herring in shallow bays at the lochside, using a technique borrowed from the salmon fishermen and still used by some of them today. At first they

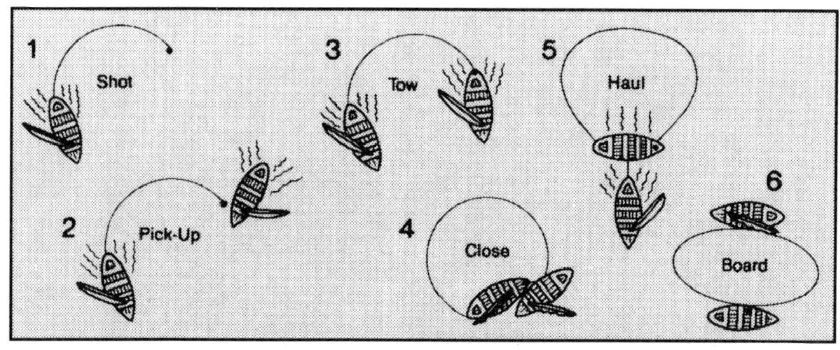

Ring-netting

Ring-net fishermen with a half-decked Loch Fyne Skiff. Tarbert, 1912.

would attach one end of a net to the shore, row out into the bay with the other end, then bring it back to the shore in a wide loop, creating a ring of net, within which, hopefully, a number of herring were caught. The net would then be hauled in, and the herring removed.

Encouraged by their success, the men began to follow the shoals in pairs of boats; one acting as the 'shore'. The net would be hauled into a loop between them and the fish removed.

The drift-net fishermen did not take kindly to this competition; partly on general grounds – they believed that the ring-netters would fish out the herring stocks, and that, in the mean time, market prices would fall – and partly because their own catches suffered by comparison. At that time, a drifter cost about £35, and a full fleet of nets the same again. In contrast, the smaller ring-net skiffs, with nets, cost no more than £20-£25 all told, and would consistently outfish the drifters.

The drift-netters had support in Parliament, and in 1851 ring-netting was banned. For ten years this Luddite law was backed by a naval presence in the Clyde, with powers to confiscate nets and use force if necessary. The fact that the fishing was generally undertaken at night made enforcement difficult, however, and the ring-netters (or 'trawlers' as they called themselves), who were largely based in Tarbert and Ardrishaig, proved difficult to deter.

In 1861 the law was repealed, and ring-netting gradually began to take over from drift-netting in the area; a process accelerated by the slow disappearance of herring from the drift-netters' heartland in upper Loch Fyne, and by the growth of the fresh fish market.

In the 18th century 70% of the Clyde herring catch was sold cured, but from that time onwards the percentage of fish sold fresh grew steadily, until it came to dominate the market.

Loch Fyne Skiffs at Campbeltown, 1900

In 1874 buyers introduced small steamers – 'herring screws' – to the Clyde, to buy the fish direct from the boats. Whenever a boat made a catch it would light a flare to attract the steamers. Each steamer which approached would send a small boat to the fishing skiff, and the fishermen would hold an impromptu auction amongst the steamer representatives. The successful steamer would then load the catch and, when it had a full cargo, steam swiftly to the nearest rail head, where the fish would be loaded onto the Glasgow trains.

Because the trawlermen lifted their nets during the night, their catches could be sold at prime prices, and were available in Glasgow by 7 am. The drift-netters, in contrast, couldn't lift their nets till dawn. Their fish were not sold in Glasgow until the afternoon, and suffered a drop in price as a result. In addition, they had to suffer the indignity of having their nets run over by the steamers, racing through the night to buy the trawlermen's catches.

The domination of the ring-net can be seen by these figures. In 1899, in the Clyde generally, 328 ring-net boats caught 55,000 crans, while 404 drifters caught only 16,000, and received less per cran as well.

A variety of types of boat were initially used by the Clyde fishermen, but by the 1880s the Loch Fyne Skiff – a double-ended sailing boat, developed from the rowing skiffs of the early ring-netters – had emerged as the ideal vessel for the fishery.

The half-decked skiff of the early 1900s was one of the prettiest sailing boats ever designed for the Scottish fishing fleets, and differed from the East Coast luggers *(see p. 26)* in a number of ways.

Firstly, it was shorter – the sailing skiffs never reached over 40', while the largest luggers were over 70' – and secondly, the skiffs had only one mast, raked back at a severe angle, while the luggers always had two. In addition, while the luggers were rigged with a dipping lug on the foremast, the skiffs always carried

Loch Fyne Skiff in regatta rig. Campbeltown, 1905

a standing lug. The standing lug is the less powerful of the two rigs, but since it does not need to be lowered every time the boat goes about, it is much handier for navigation in confined waters. This need for manoeuvrability – both when sailing in general, and when handling the ring-net in particular – also dictated the underwater shape of the boats, which were deeper at the stern than the bows, thus making them more responsive to the helm.

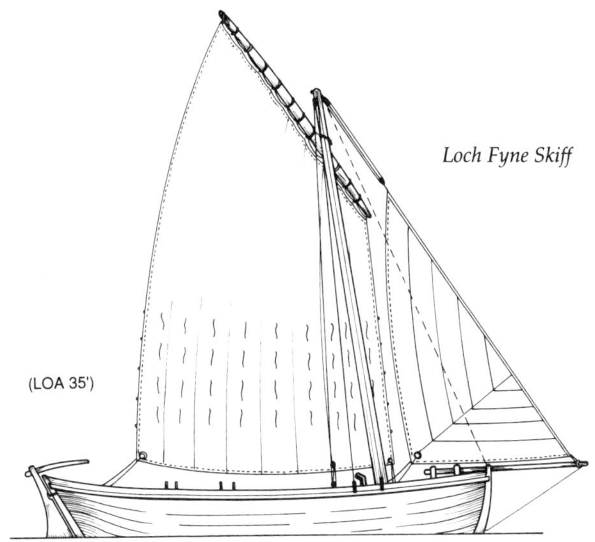

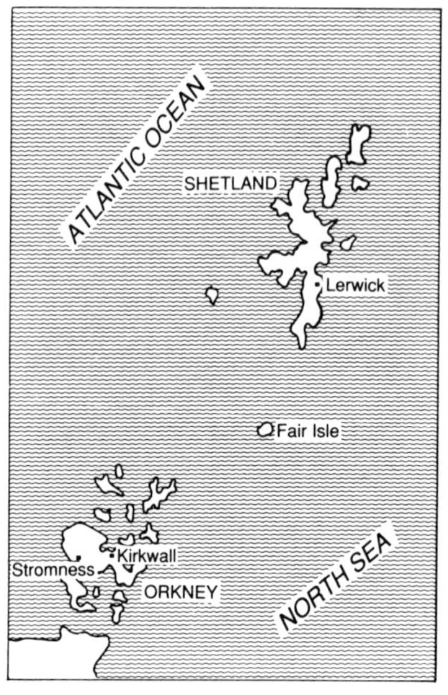

THE WATERS around the archipelagos of Orkney and Shetland, off the north of Scotland, are very rich in fish. The men of both island groups were involved in fishing from an early period, but, despite their proximity to Caithness, the Orkney Islands did not attract a large-scale herring fishery; the people of the islands tending to look upon agriculture as their primary source of income.

In the rather less fertile islands of Shetland these priorities were reversed, and the herring fishery did have a large impact. It arrived comparatively late, however, and achieved primacy in the fishermen's lives only by eclipsing an older, indigenous industry which had a number of unique features.

Shetland only became a part of Scotland in 1469, and the islands were still a distant and largely forgotten corner of the United Kingdom as late as the early 19th century. The people – who were principally of Scandinavian descent, and still spoke a Norse dialect called Norn – were the poor tenants of a small group of Scottish landlords, whose major financial interest in the islands lay in the annual line-fishing for ling.

Shetland boats in Lerwick

Haaf Station at Stenness, 1880s

The isolation of the islands gave the landlords a firm control of every aspect of the industry: importing equipment, curing and selling. This control enabled them to develop a system whereby their incomes were virtually assured, since they did not pay the fishermen until after the cured fish had been sold, and the value of the catch was known. The fishermen's pay fluctuated accordingly.

Since their primary income came from fishing, the landlords attempted to make sure that their tenants took an active part in it; going so far as to make their continuing tenancy dependant on their participation. The major fishery – great-lining for ling – was called the 'haaf', and took place between May and August, with the fishermen gathering at the stations established by their respective landlords. These stations consisted of a series of small huts situated on the outer edge of the islands, as near the fishing grounds as possible. They were placed behind broad, stony beaches, suitable for drying the fish. A large station might consist of 70 boats, with accommodation for 400-500 fishermen and curers.

The fishermen would row or sail their boats up to 40 miles before setting their lines. These could stretch up to 6000 fathoms (that is, over seven miles) along the sea-bed, and carry between 1000-2000 hooks, baited with herring, haddock or saithe. The routine of setting these lines, laying to them for two to three hours, then hauling them, could take between six and eight hours, and the routine might be repeated two or three times before returning to the fishing station. Trips were undertaken once or twice in a week.

Incredibly, this most dangerous fishery was carried out in small, open boats; albeit they were built to a design of the finest pedigree.

The Vikings colonised Shetland in the 8th and 9th centuries, and they brought with them both a tradition of great seamanship and their long, narrow, double-ended, clinker-built ships. The lack of trees on Shetland made boat-building impossible, but during the following centuries the islanders maintained their links with Scandinavia, and imported Norwegian boats, or pre-shaped planks suitable for their construction, both for fishing and for communicating between the islands.

The Shetlanders could make particular design requests of the boat-builders – to produce vessels more suitable for their own peculiar requirements – but until the middle of the 19th century the Shetland boats remained very close to the Norwegian models. From then on, however, the links with the Scottish mainland began to become more regular, and the islanders started to import wood from Scotland and to build boats to their own designs.

The general form of the vessels remained close to that of the Scandinavian boats, but a number of different classes evolved to suit the requirements of the different types of fishing. Broadly, these were the sixereen, the fourereen and the yoal.

The sixareen (or sixern) was the largest of the Shetland boats; a six-oared boat up to 36' in length and used for the haaf fishing. The fourereen (or fourern) was a smaller version; a four-oared boat up to 26' in length, and used for the winter haddock fishing. The yoal was the smallest and the most elegant of the three: a low, narrow boat of about 22' in length, used for catching saithe by hand-line; rowing alongside the furious tide rips between the islands.

The basic form of all these boats was similar. They were light, so that they could be hauled up the beaches, but of a very seaworthy design. They were narrow at the waterline and low in the waist, to make rowing fast and easy, yet they were high-ended, with flaring topsides and long overhang, providing a lot of potential buoyancy, if a large haul demanded it. The boats were designed to be either rowed or sailed, and a short, stayed mast could be set virtually in the centre of the boat. At first, these masts carried square sails, like Viking ships, but although this tradition persisted in Fair Isle, most of the Shetland boats adopted a low dipping-lug.

The Shetland boats were notably fast under both oar and sail, and though they

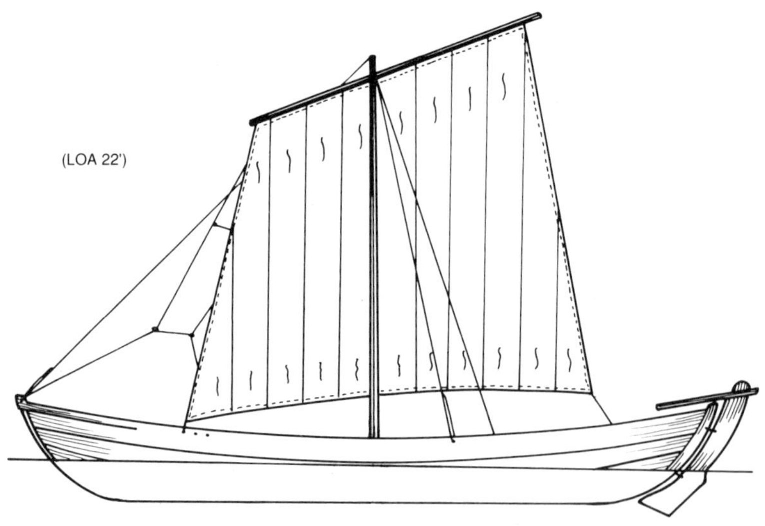

Shetland Yoal

Lerwick in the herring season, 1890s

appeared slight and somewhat dangerous, it was repeatedly proved that, given a good crew, they were capable of riding out almost any storm.

Around 1880 the first curers began to arrive in Shetland, establishing herring bases throughout the islands along the model of those used in the Western Isles. Within a short time the haaf fishing had been largely neglected, and the bulky sixareens with it. The smaller boats for inshore fishing were retained, however, and their descendants can still be seen in the islands today.

After having been so closely tied to the landlords, the Shetlanders took keenly to the freedom of the herring fishery, with its potential for large personal profit. By the end of the century there were approaching 400 island-owned fishing luggers in the fleet of around 1400 which gathered for the Shetland fishing.

Herring were caught from May to June off the northern islands of the group, and from July to September off Lerwick and the southern islands; thus, for most fishermen, the Shetland season was not an addition to the East Coast and Minch fisheries, but an alternative.

This long home season provided a fine opportunity for the Shetlanders to prosper, and although the local industry suffered economic problems similar to those experienced in the Western Isles, they were not quite so acute. Nonetheless, there was a considerable contraction of the fleet when steam replaced sail.

The herring boats used by the Shetlanders were generally similar to those found elsewhere in Scotland, but most of the islanders abandoned the dipping lug and adopted the gaff ketch (or 'dandy') rig used by the English East Coast fishermen.

'Dandy' rig

A Zulu and a Fifie entering Great Yarmouth harbour

With the exception of the standing lug on the Loch Fyne Skiff, and the ketch rig adopted by most of the Shetland fishermen, the rig which dominated the Scottish herring fleet was the dipping lug. At one time this rig was used on many of the work boats around the British coast, but in most places it was gradually replaced by the handier gaff rig, until only the Scots and a few others retained it.

The dipping lug is a simple rig, with a great many virtues and one serious vice. As the diagram shows, it is essentially a square sail set asymmetrically on the mast, with a raised peak and the tack attached inboard. It is a cheap rig, with only two ropes (a halyard and a sheet) and little which is likely to break. It sets its canvas low down, so the vessel is less likely to be overpowered in a breeze than a gaff rigged boat. Above all, the dipping lug is a powerful sail, and in an industry which depended on speed this was a great advantage.

The drawback of the rig is that when tacking (turning the boat through the wind when working to windward), the sail has to be dropped at each turn, lifted round the mast, and hoisted again on the new tack. On a large lugger this operation could take about ten minutes, and it required the six man crew working in skilful concert to complete the operation safely. This was a somewhat dangerous procedure in a breeze, and rendered the boat awkward in confined spaces.

Nevertheless, the Scots fishermen liked the rig, and far from seeking to abandon it, they continued to develop bigger and faster luggers right up to the end of the sailing era.

The boats on which these lugsails were set were as simple, cheap, powerful and elegant as the rig which drove them. They were developed from the small open boats of the East Coast, which in turn may have been descended from the Viking ships. At any rate, the Scottish fishing vessels shared a number of features with the longships, being double-ended, long-keeled and (initially) clinker built.

The first accurate illustrations which emerge are those which accompanied the report compiled after the disastrous storm of 1848. The boats were undecked, about 40' long, and built to a variety of designs.

The report was generally scathing about both the boats and their rigs, but the boatbuilders and fishermen defended them strenuously. The nature of the industry, coastline, and local finances dictated the type of boat in use. It had to be light and of a shallow draught (for when it was operated off the shore), fast (in order to bring the herring to the curers as swiftly as possible), compact (for lying in crowded harbours) and, above all, cheap. The stripped-down functionalism of the designs, even at the expense of the safety and comfort of the men, was to distinguish the Scottish herring luggers right up to the end – even the largest luggers never had bulwarks fitted around the deck: an advantage when handling the nets, but dangerous for the men in rough weather. It was largely through such single-mindedness that the Scots came to dominate the British herring industry.

Such improvements as were made evolved slowly as the market dictated and as changing conditions allowed. Potential demand for herring outstripped supply for most of the 19th century, so there was continuous pressure to catch more fish. The individual fisherman could best achieve this by using a larger boat. At first this was impractical, due to the lack of harbour space, but as the size and number of harbours increased, so did the size of the boats. The larger the boat, the further it could travel in search of herring; thus it could fish for longer each year and the fisherman's income increased accordingly. After a few seasons

A Berwick Fifie and crew

he would buy a still larger boat.

Two inventions of the late l9th century had a dramatic effect on the size of the herring boats: cotton nets and the steam capstan.

Nets had traditionally been made of hemp or flax, but the new cotton yarn, introduced in 1860, was only one third the bulk of the older materials, and proved to be twice as durable. Fishermen could now carry a greater expanse of nets, and they soon discovered that their boats were too small to carry the increased amount of fish which it became possible to catch. Larger boats were built; from the 1860s onwards fully-decked.

The steam capstan (or winch) was introduced in the 1880s, and performed two functions onboard the luggers. Firstly, it was used to help hauling the nets. This allowed an even larger number of nets to be set (up to 70 ultimately, creating a total net-area of over 40,000 square yards), and thus further increased the catch of each boat. Secondly, it was used to hoist the great lug sails. Since the ultimate rig-size of the fishing boats had previously been dictated by the amount of canvas which the crew could hoist aloft, this invention greatly increased the potential size of the sails, and thus of the boats.

Under the impact of these two changes, the luggers finally reached a length of just under 80': the largest luggers ever built.

In 1848 the hull forms of the double-ended luggers varied considerably from area to area, but already two types were emerging as the most popular. On the East Coast, south of Fraserburgh, the fishermen used a narrow boat with a virtually straight stem and stern and a long keel. This design – the 'Fifie' – was fast and powerful, but its long keel rendered it somewhat slow in response to the helm, and thus rather unmanoeuvrable.

The 'Scaffie', used off the coast of the southern Moray Firth, had a greater beam and a shorter keel, with the stem cutting away sharply beneath the

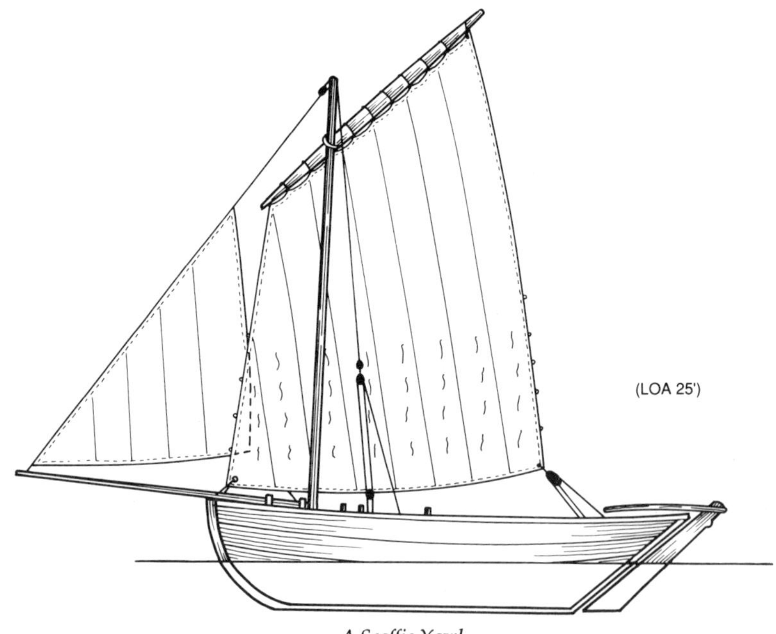

A Scaffie Yawl

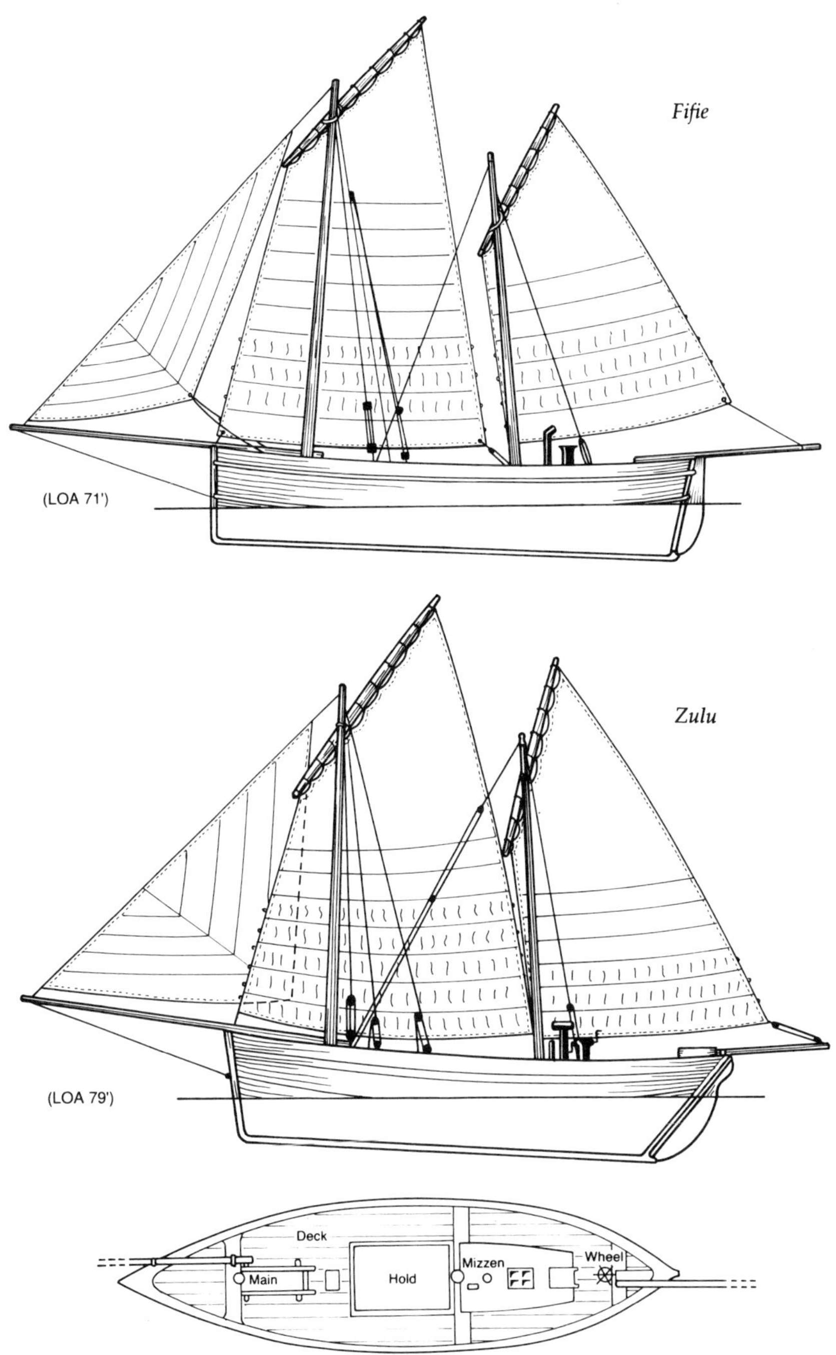

Fifie
(LOA 71')
Zulu
(LOA 79')
Deck
Main
Hold
Mizzen
Wheel

Fully-rigged Zulu

waterline, and a long, straight, angled stern. This was a more seaworthy design, but lacked the speed of the Fifie.

As the boats became larger, and a seaworthy shape less vital, the Scaffie sailors began to look for a design which would match the speed of the Fifies. In 1879 a boat called the *Nonesuch* was built and launched at Lossiemouth, incorporating the straight, deep stem of the Fifie and the angled stern of the Scaffie. This design – called a 'Zulu', after the war which was underway at the time – was felt to have the best qualities of the two older types, and quickly superseded the Scaffie along that stretch of coast.

In general terms, the Fifie was built and used by the fishermen of the East Coast, between Peterhead and Berwick, plus those of Caithness and east Sutherland and the Northern Isles, while the Zulu dominated on the southern coast of the Moray Firth, and also in the Western Isles.

Fully-rigged Fifie

*The Conclusion*______________________

With the exception of a few bad years, the herring industry grew steadily throughout most of the 19th century. In the 1830s, the average production was around 400,000 barrels per year, while by the 1880s this figure had reached around 1,500,000 barrels. The whole industry was geared for continuing expansion, and financed by extended credit based on the curers' guarantee of fish prices to their tied crews.

In 1884 the market collapsed. The vital German market was flooded with a large quantity of low quality herring and the resulting prices were too low to balance the curers' books. Over the following three years the quality of the catch failed to improve, and ultimately the banks took fright and foreclosed on a number of curers.

After ten years the industry began to pick up again, but with one vital difference: the guaranteed price system was replaced by auctions.

Auctions were not an entirely new idea, having been used in Fife for some time. There was resistance to them amongst the fishermen in the North, however; particularly those fishing from quieter coasts, where a shortage of competing curers could severely reduce prices. As a result, when auctions became general, the herring industry became more and more concentrated on a small number of busy ports where the curers tended to congregate. By 1913, 89% of the East Coast catch was landed at Wick, Fraserburgh, Peterhead or Aberdeen.

Meanwhile, large numbers of Scottish boats – always searching for ways to extend their fishing season - began to pursue the herring shoals which gathered off East Anglia in the autumn.

The Fife boats had begun making the trip in the 1860s, but it was not until the 1890s that the other areas joined in; lifting the number of Scottish boats involved from 200-300 in the 1860s to over 1000 in the early years of the 20th century.

Yarmouth and Lowestoft were the main ports for the visiting boats, and between October and December they were dominated by the Scots; with the harbours full of the distinctive luggers, and the curing yards largely operated by the Scots herring girls, who followed the fleet each year and lived in cramped lodgings in the towns for the duration of the fishery.

Around the turn of the century the first steam drifters were introduced into the Scottish fleet, and within a few years the building of new sailing luggers had virtually ceased. The cost of the steam drifters was prohibitive to the poorer fishermen, and luggers continued to be sailed for as long as the old boats were seaworthy – into the 1930s in some outlying areas – but the trend was irreversible, and in most areas the change was swift.

For those who could afford them, the arguments in favour of the steam boats were convincing. They might be expensive to buy and operate, and slower than the luggers when the wind was kind, but they were untroubled by calms, and over a season they could outfish any sailing vessel. In 1910 over 1200 Scottish boats made the trip to East Anglia, of which over half were steam drifters. Over the season the steam boats made an average of £530, while the luggers averaged only £134. In fact, once the expenses of running the steam drifters were taken into consideration, the incomes of the steam and sail crews were not that different, but they were different enough to ensure the end of the Scottish sailing drifters.

Through all these changes the industry continued to grow, and any expenses were absorbed by increases in catches, prices and sales. Like so much, however, the Scottish herring industry was to be destroyed by the horrors of the First World War.

The four years of warfare were bad enough; destroying the carefully established European export chains through which 80% of Scottish herring were sold and strangling investment. Worse, however, was the subsequent loss of markets, with the huge Russian trade disappearing with the advent of the communists, and the poverty-stricken Germans lurching from chaos to disaster.

Following World War II there was a recovery in the fortunes of the Scottish fishing industry, but although the trade remains important to this day, it has never regained the glamour and excitement which it had in the days of the great sailing luggers.

Bibliography

The following books are invaluable to anyone with an interest in Scotland's sailing fishermen, and contain a great deal of information on subjects which can only be touched upon in this publication. The titles are self-explanatory with the exception of 'The Silver Darlings', Gunn's fine novel which uses the growth of the Caithness herring fishery as a background.

'Sailing Drifters' Edgar J March (1969)
'The Ring-Net Fishermen' Angus Martin (1981)
'The Silver Darlings' (novel) Neil M Gunn (1941)
'The Fishing Industries of Scotland: 1790-1914' Malcolm Gray (1978)
'Scottish Sail' Robert Simper (1974)
'The Sixareen and her Racing Descendants' Charles Sandison (1981)
'The Shetland Boat: South Mainland and Fair Isle' Adrian G Osler